Birds of Prey

HAWKS

LYDIA LUKIDIS

BLACK RABBIT BOOKS

Bolt is published by Black Rabbit Books
P.O. Box 227, Mankato, Minnesota, 56002
www.blackrabbitbooks.com

Marysa Storm, editor; Michael Sellner, designer and photo researcher

Library of Congress Cataloging-in-Publication Data
Names: Lukidis, Lydia, author.
Title: Hawks / by Lydia Lukidis.
Description: Mankato, Minnesota : Black Rabbit Books, [2023] | Series: Bolt. birds of prey | Includes bibliographical references and index. | Audience: Ages 8-12 | Audience: Grades 4-6 | Summary: "Equipped with strong senses and powerful features, hawks rule the skies. They also capture the imagination of readers everywhere. With an engaging design focused around expertly leveled text and colorful infographics that support visual literacy, introduce reluctant and beginning readers to the life cycle, features, and diets of these powerful birds of prey" Provided by publisher.
Identifiers: LCCN 2020018010 (print) | LCCN 2020018011 (ebook) | ISBN 9781623105549 (hardcover) | ISBN 9781644664858 (paperback) | ISBN 9781623105600 (ebook)
Subjects: LCSH: Hawks–Juvenile literature.
Classification: LCC QL696.F32 L845 2023 (print) | LCC QL696.F32 (ebook) | DDC 598.9/44-dc23
LC record available at https://lccn.loc.gov/2020018010
LC ebook record available at https://lccn.loc.gov/2020018011

Image Credits

123RF: cedecea, 16; deniskrivoy, 6–7, 14–15; Alamy: All Canada Photos, 23; Beata Predko, 27; Chris Howes/ Wild Places Photography, 4–5; Genevieve Vallee, 20; Jim Cumming, Cover; Johann Schumacher, 23; Murray Hayward, 23; Ondrej Prosicky, 16; Robert Kennett, 16; Tariq Hameed Sulemani, 11; Dreamstime: Le Thuy Do, 5; Mcasabar, 23; Ondrej Prosicky, 16; Robert Hambley, 29; Stephen Mcsweeny, 28–29; Wildlife World, 12–13; Getty: Per Grunditz / EyeEm, 4; Minden Pictures: Bill Baston, 15; Eric Medard, 11; iStock: Murphy_Shewchuk, 21; Shutterstock: Chase D'animulls, 11; Edward R, 32; Eric Isselee, 8–9, 25; Feng Yu, 25; IgorCheri, 25; JayPierstorff, 1, 31; Jiri Fejkl, 15; JRP Studio, 23; Le Do, 25; Lumena, 18–19, 27; Mahi ryan, 3; mariamalaya, 21; MTKhaled mahmud, 11; Richard G Smith, 25; Tony Campbell, 22; Wirestock Creators, 6

Contents

CHAPTER 1

STRONG Hunters

The sun begins to set. A large group of bats leaves a cave. They're on the hunt. But so is the red-tailed hawk flying above them.

The hungry hawk follows the bats. It watches as a bat flies away from the group. The hawk has found its **prey**.

Swiftly, the hawk swoops down. Its sharp talons grab the bat. Time to eat!

COMPARING WINGSPANS

SHARP-SHINNED HAWK

17 to 28 INCHES
(43 to 71 centimeters)

Different Types

Hawks are powerful birds of prey. These **predators** are often small or medium in size. Most have short, rounded wings that help them soar. Hawks' beaks are short and hooked. Their strong feet have talons. They use these talons to catch and kill prey.

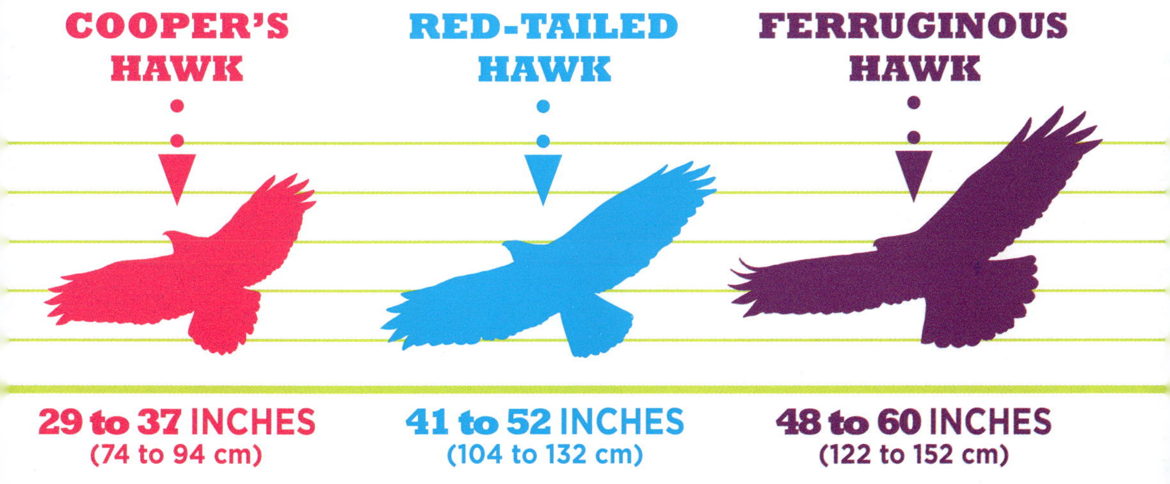

PARTS OF A HAWK

Four Groups

There are four main groups of hawks. Each kind looks a bit different. Buteos are the biggest type. They have short, wide tails. Accipiters have long tails that help them change direction while flying. Meanwhile, kites have shorter legs than other hawks. They also have long, pointed wings. Harriers have thin bodies and long legs. Their faces are flat like an owl's face.

BUTEO

ACCIPITER

KITE

HARRIER

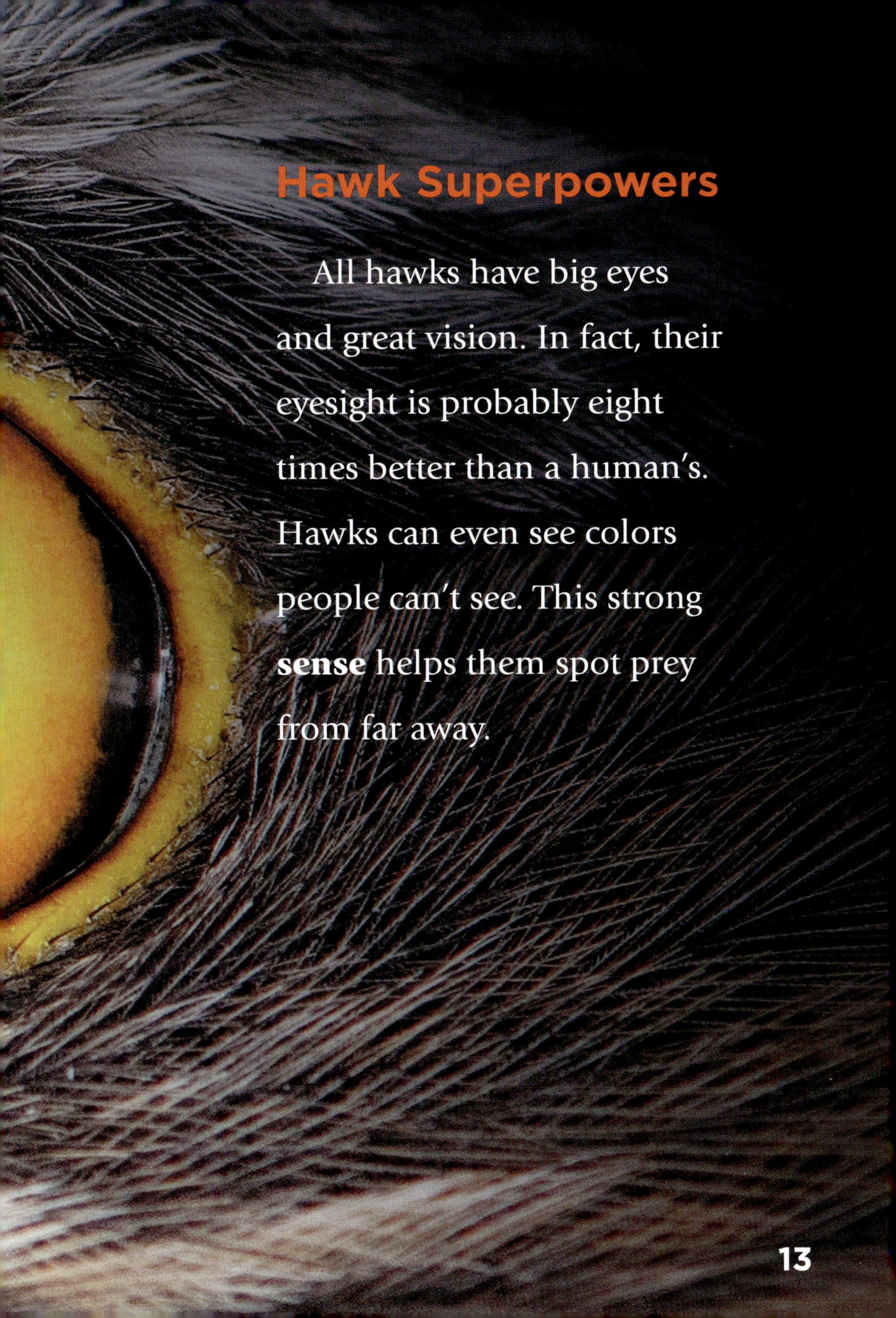

Hawk Superpowers

All hawks have big eyes and great vision. In fact, their eyesight is probably eight times better than a human's. Hawks can even see colors people can't see. This strong **sense** helps them spot prey from far away.

CHAPTER 2

At HOME and on the Hunt

Hawks eat food they catch. Many hawks hunt animals that fly in the sky. They often chow down on bats and smaller birds. Some hawks also hunt animals on the ground, such as **reptiles** and small **mammals**.

Hawks often hunt from perches, such as trees and telephone poles. From the perches, they look for prey. Once they spot prey, they attack.

Hawks will also look for prey while soaring above fields. When they see a small animal below, they swoop down.

FORESTS

GRASSLANDS

DESERTS

Habitats

People often see hawks. The birds' populations are growing. And they live all around the world. They often fly through forests, grasslands, and deserts.

Hawks are superfast. The red-tailed hawk can go more than 100 miles (161 kilometers) per hour when diving.

HAWK RANGE MAP

Hawks live on every continent except Antarctica.

ASIA
EUROPE
AFRICA
AUSTRALIA
ANTARCTICA

CHAPTER 3

HAWK Families

Most hawks fly and hunt alone. They only meet to have babies. Often, a hawk picks the same **mate** each year. The pair works together to make their nest using twigs and bark. These nests are usually high in trees. A pair sometimes uses the same nest each year.

Where Hawks Nest

CACTI

Baby Hawks

Once the nest is ready, the female hawk lays eggs. About a month later, the eggs hatch. At first, chicks need their parents to feed and protect them. The chicks quickly grow stronger, though. After about six weeks, young hawks begin to fly. It's soon time to leave the nest and care for themselves.

Comparing Average Clutch Size

red-tailed hawk | Cooper's hawk | ferruginous hawk | sharp-shinned hawk

CHAPTER 4

PREDATORS and Other Dangers

Big hawks face few predators, but small hawks and chicks must watch out. Raccoons and other mammals go after nests. Some hawks eat smaller hawks. Hawks steal food from each other too.

Hawk Food Chain

This food chain shows what eats hawks.
It also shows what hawks eat.

Human Threats

Hawks of all sizes must watch out for humans. Some people hunt hawks. They also turn hawk habitats into fields for farming. They cut down forests for wood. Hawks have fewer places to live and hunt.

Protecting hawks is important. Some hawks are food for other animals. And they all help keep **pest** populations down. These powerful birds play a big role in the world.

In the 1940s, people made a poison called DDT. They used DDT to keep pests from crops, but it killed many hawks. The poison is now illegal in the United States.

By the NUMBERS

2 to 4.5 POUNDS (1 to 2 KILOGRAMS)
weight of a ferruginous hawk

1972 YEAR THE UNITED STATES MADE DDT ILLEGAL

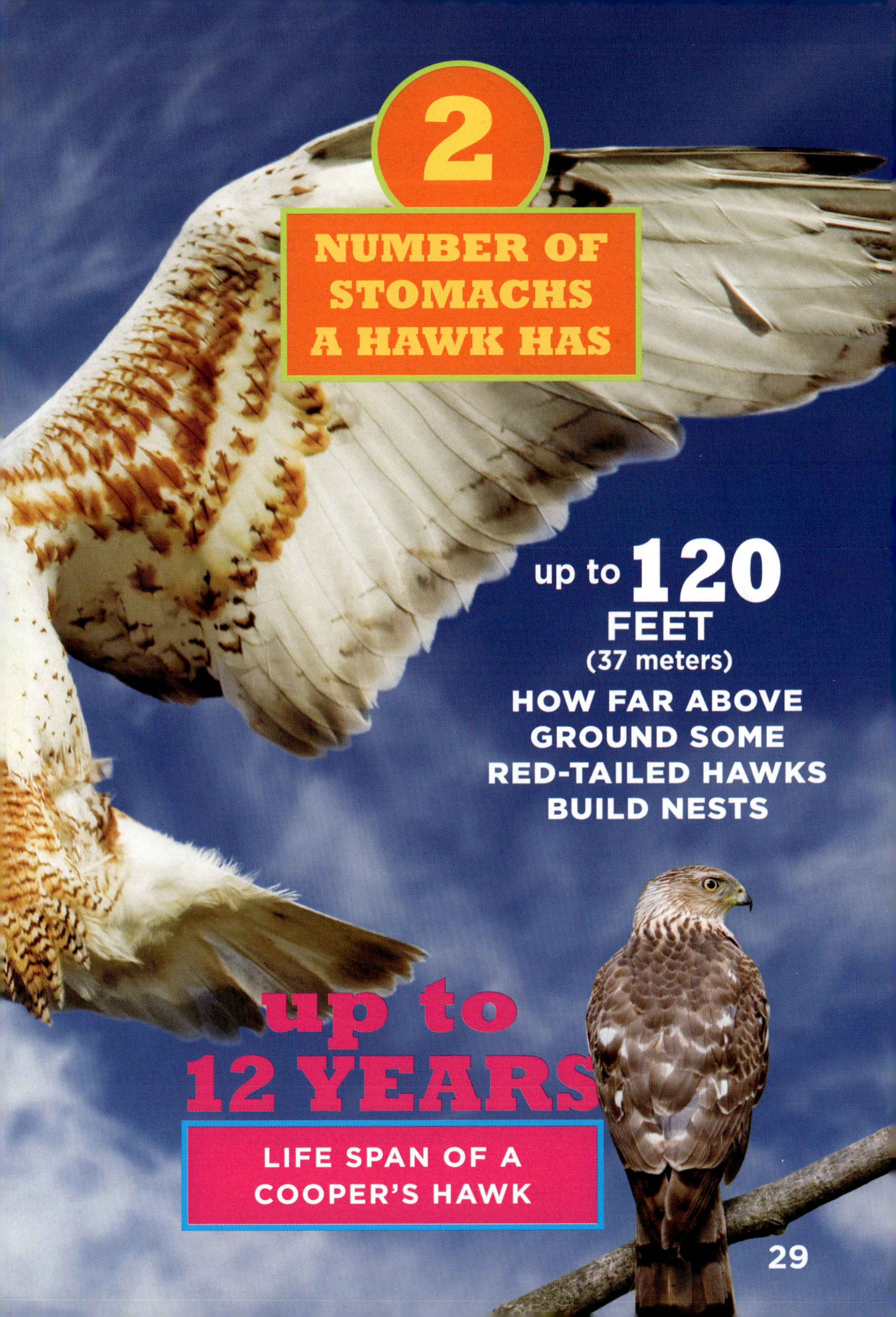

2

NUMBER OF STOMACHS A HAWK HAS

up to 120 FEET

(37 meters)

HOW FAR ABOVE GROUND SOME RED-TAILED HAWKS BUILD NESTS

up to 12 YEARS

LIFE SPAN OF A COOPER'S HAWK

mammal (MAH-muhl)—a type of animal that feeds milk to its young and usually has fur

mate (MAYT)—one of a pair that joins together to produce young

pest (PEST)—an animal or insect that causes problems for people especially by damaging crops

predator (PRED-uh-tuhr)—an animal that eats other animals

prey (PRAY)—an animal hunted or killed for food

reptile (REP-tahyl)—a cold-blooded animal that breathes air and has a backbone; most reptiles lay eggs and have scaly skin.

sense (SENS)—one of five natural powers from which you receive information; the five senses are touch, taste, smell, sight, and hearing.

BOOKS

Gleisner, Jenna Lee. *Hawks.* Raptor World. Minneapolis: Jump!, Inc., 2020.

Riggs, Kate. *Hawks.* Amazing Animals. Mankato, MN: Creative Education, 2022.

Walker, Alan. *Birds of Prey.* Built to Survive. New York: Crabtree Publishing, 2022.

WEBSITES

Hawk
kids.britannica.com/kids/article/hawk/353234

Red-Shouldered Hawk
www.nhnature.org/visit/animal_info_sheets/redshouldered_hawk.php

Red-Tailed Hawk
kids.nationalgeographic.com/animals/birds/facts/red-tailed-hawk

INDEX